My Parents Second Computer and Internet Guide. *Beyond the Basics*

"Live, Love, Learn"

The sale of this book, and other books published by KLMK Enterprises, help raise funds for cancer research. Find a cure.

www.myparentsfirst.com

Published by KLMK Enterprises, www.myparentsfirst.com
An Enterprising Publishing Company
Printed in Canada

ISBN 978-0-9732728-2-6

2007

Beyond the Basics

Before you go beyond the basics, feel comfortable using your computer, learn at your own pace.

Know the basics, found in *My Parents First Computer and Internet Guide.*

You will find I am more direct in this book, for example:

Move your mouse to Start,
up to Programs, then to Accessories
and finally to Paint. Left-click

Will read simply as:

Start > Programs > Accessories > Paint

By using the ">" sign I am assuming you have learned how the mouse works and know the basics of looking for and opening a program, folder or file.

Ready?

Let's go!

Beyond the Basics

Customizing Your Computer

Revving Up Word

Beyond the Basics

Excel in a Nutshell

Internet Savvy

Customize your Computer

Customize your Computer

Customizing Your Computer

Inside the *Control Panel* you will find tools that customize and help configure your computer.

On the next few pages I'll show you how to use the most commonly accessed tools: Display, Printers & Faxes, and Add/Remove Programs. Take the time to notice what else is in there - *like, changing the Date and Time...*

- **Display** (What your desktop looks like)
- **Printers & Faxes** (to add a printer)
- **Add/Remove Programs**

To Enter the Control Panel:

Start > Settings > Control Panel

My Computer, My Way! Yeah!

Display

Display Properties of your Desktop

"Display Properties" Simply means what your desktop looks like and how it acts.

What's a "Desktop?
The desktop is what your monitor looks like when you don't have any programs running.

You can personalize it with your own pictures, change the size of the icons, change your screen saver...
All sorts of things!

There are two ways to open the Display folder:

A. Settings > Control Panel > Display

Or the easy way!

B. Right-click anywhere on your desktop and choose *Properties*

Display Properties				
Themes	Desktop	ScreenSavers	Appearance	Settings

Display

With Windows XP, Display Properties opens on a tab showing what your current **Theme** looks like.

Themes are the "*full meal deal*" in the way that they offer pre-set formats for all the other properties!

Picking a Theme is easy, but I like my desktop personalized with my favorite pictures. Check out the other tabs, experiment a bit and personalize yours. Here's how!

The other tabs are:

Desktop, Screen Saver, Appearance & Settings.

Display Properties				
Themes	Desktop	ScreenSavers	Appearance	Settings

Desktop, here's how to get there:

1. Display Properties > Desktop tab
2. Click on *Browse* to open a path and search for a saved picture in your computer, or use one of the ones already there.
3. Highlight the picture's name, click OK.

That's it!

Windows XP also offers a great feature for personalizing your desktop. **Right-click** directly over a picture you have open and see an option to place the picture on your desktop as wallpaper!

It can't get much easier than that!

Display

Display Properties				
Themes	Desktop	ScreenSavers	Appearance	Settings

Screen Savers
Screen savers originated to prevent *ghosting,* which happened with early-technology monitors. Ghosting is almost non-existent with current technology, so today's screen savers are mostly decorative and entertaining.

Virtual fish or a crackling fire are popular screen savers. I like my screen saver to scroll through my favorite family photos!

Screen savers can also offer a certain level of privacy, by requiring a password to re-open your system once active.

Here's how to change your Screen Saver:

1. Display Properties > Screen Saver tab.

2. To open a list of installed screen savers, click on the arrow beside the Screen Saver name box.

3. Click on the Screen Saver of your choice.

4. Click on OK or Apply.

Display

Display Properties				
Themes	Desktop	ScreenSavers	Appearance	Settings

Appearance
Have a look at the color and font in the Task Bar (beside the Start menu button) or along the Title Bar at the top. You can change the color and print to almost anything you desire!

Here's the how and what of Appearance.

I find that the headings under Appearance are a little confusing. Here's the scoop on what they do:

- *Windows & Buttons,* can alter the style of the bars.
- *Color Scheme,* changes the color in the bars.
- *Font Size,* changes the size of type by the icons on your desktop from small to large.

Click *OK* or *Apply* to accept the changes.

Tip

If you're not sure about making changes, just jot down what the settings are before you alter them. So you can easily change them back!

Display

Display Properties				
Themes	Desktop	ScreenSavers	Appearance	Settings

Within *Display Settings*, you can adjust the screen resolution or the spectrum of colors that appear.

Here's what Screen Resolution means:

Pixels - A pixel is a dot. The display on your screen is made of up these dots (pixels). By the way, your printer prints in pixels too!

A 15-inch monitor might display 640 pixels on each line and have 480 lines, making it a 640 x 480 pixel screen. That equals to about 50 dots per inch (dpi) and you'll see a pretty clear picture.

A higher resolution screen will be able to show more pixels per inch, with a clearer picture. My monitor works with 96dpi, but is capable of 120dpi.

Why would you ever adjust these? There are some programs that will only run with certain resolution and color settings.

If you decide to change your display settings, I recommend that you write down what the original settings are, before you muck with them!

Printers and Faxes

Your printer is one of those things that should just always work. But just like my car, trouble always seems to pop up at the worst time. Luckily, you don't have to be a mechanic to fix your printer!

Settings>Control Panel> Printers & Faxes

Installing & trouble shooting your printer
Once you know where to go and what your computer is looking for, installing your printer and other devices is pretty straightforward.

Each device attached to your computer is driven by a *Driver*, including your printer. You don't really have to know much about drivers, but a wee bit of knowledge here helps things make more sense.

Here's the skinny on Drivers!
The main chip in your computer speaks in a very generic language. Remember about protocols that the internet uses? This is the same kind of thing.

Every device speaks in its own secret language, I guess to protect *brand* technology. So, between the device and your computer, a translator so-to-speak, is needed. *Hence the Driver.*

Bonjour = Hello
Hello = Bonjour

The Driver for your printer is part of its installation disk. If you do not have an installation disk you might have to download a driver from the maker's web site.

Printers and Faxes

Installing & Trouble Shooting a Printer

Here's how:

1. Start > Settings > Control Panel > Printers & Faxes
2. The Printers and Faxes window will open to show you few options, including "Add a Printer". It will also show you if you have any printers already installed.
3. If you already have a printer installed it will name the printer. If it's active, it will have a ✓check mark beside it.

If you're adding (installing) a printer, click on that option and just follow along the prompts.

If you're having trouble with your printer, you will see its name. If its active, it should show a check beside it. If it's not "active", maybe that's the only problem!

- Click on, and open your active printer. A small window will pop up showing any documents in queue (waiting to be printed).
- Click on Printer>Properties. You will see an option to "Print a test page". Try that. If it doesn't work, your program will then try and help you fix it, by walking you through its *trouble shooting* steps. Good luck!

Add/Remove Programs

Add/Remove Programs

When we got our first computer, our *computer guru Pat* couldn't stress to us enough that when it came to **removing programs,** to **use this feature**. Over the last decade I have come to understand why.

You can easily delete a program simply by right-clicking on its name, found through the start menu. But if you delete this way, you will only delete what you see, and leave crumbs of it all around in your system.

Using *Add/Remove Programs* is like having Super-maid help with the housework. Every little part of the program gets picked up, cleaned out and removed, even the crumbs!

Here's where:

Settings > Control Panel > Add/Remove Programs

The Add/Remove window will open to display a list of all the *Programs* that are installed in your computer. If you click over the name of a program, you will see options to Change or Remove it.

Choose what you want and *just do it!*

I've never used the Add Program option because most programs know how and where to load themselves, including Windows Updates.

General Maintenance

System Tools

The *System Tools* are amazing. It's almost like having a computer doctor right there with you all the time. Some of the tools can even repair the problems they find!

A little regular maintenance will help your computer run smoothly. These tools are easy to start, and they do all the work by themselves! **You** just have to remember to use them!

If you notice your computer is running slowly, running Disk Cleanup and Disk Defragmenter might help fix the problem. You can even schedule them to work on their own!

We will look at these tools on the next few pages:

Disk Cleanup
Disk Defragmenter
Scheduled Tasks

There are other tools in there. When you get a chance, have a look at them on your own.

General Maintenance

Disk Cleanup

Disk Cleanup can help free up space and clutter in your hard drive. It looks for things like temporary internet files and program files that you might not need anymore, and feels you can safely delete.

After you run Disk Cleanup, it will ask if *you are sure you want to delete these items.* You can view them if you want. I usually say *Yes* to delete them and have never regretted it.

Here's how:

1. Start > Programs > Accessories > System Tools > Disk Cleanup
2. Choose where in your computer you want Disk Cleanup to work. The "C" drive is your main drive and the most important to do.
3. When the job is complete a report will pop up showing you the results.
4. You can choose to view the files it has suggested you delete, or just click OK and be done with them.

TIP:

Maintenance programs will run faster if all other programs are turned off.

General Maintenance

The Disk Defragmenter!

Just like *Mr. Spock,* computers want things in a logical order. The Disk Defragmenter helps to keep files organized by picking up fragments of data that was misfiled, and putting it where it belongs.

Here's how:

1. Start > Programs > Accessories > System Tools > Disk Defragmenter

2. Click *Defragment*

   Yep, that's it.

Something to know about the defragging, **it is not quick!** The messier your computer, the longer it's going to take...

Remember, System Tools work faster if you are not running any other programs.

General Maintenance

Scheduled Tasks

This little wizard can help you set up a regular maintenance program. *Scheduled Tasks* can also get your computer to do a bunch of jobs automatically, even check for e-mail!

Open it to see a whole list of programs, it will then ask you what programs you want to run and when.

Here's how:

1. Start > Programs > Accessories > System Tools > Scheduled Tasks
2. A *Wizard* window will pop up asking you what programs you would like to run.
3. Follow the directions, step-by-step, with the help of the *Wizard*.
4. Click Finish when you're done.

Tip:

For Scheduled Tasks to work, your computer must be on during the times you have told it to run!

Deleting Files

You've probably discovered that it is easy to create files by mistake, I know I do. Don't worry, it's easy to delete files or folders.

Here's how to delete a file in Microsoft Word:

1. Open *Word*
2. Click on: File > Open

A window similar to this one will pop up.

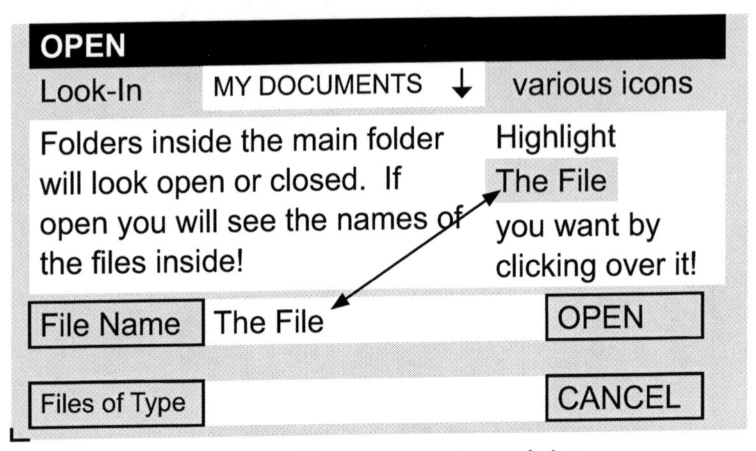

3. Highlight the file you want to delete
4. **Right-click, choose Delete, left-click.**

Select
Open
Print
Copy
Create Shortcut
Delete
Rename
Properties

Trash and Recycling

Your computer has a Trash Can and/or a Recycling Bin. When you delete a file or folder, it goes directly there.

Here's where:
You should have an icon on your desktop for your trash or recycling, left-click on the icon to open it.

When your Trash/Recycling folder is open, it will show all the files and folders that you have deleted since the last time it was emptied. This is a good thing, because if you deleted something by mistake, at this point it's easy to get it back.

Notice the options in the left sidebar:

- Empty Recycle Bin
- Restore all items

If you highlight a single file,
"Restore all items"
will change to
"Restore item"

The Recycling Bin is your safety net!

Trash and Recycling

To Retrieve from the Recycle Bin

1. Open the Recycle bin
2. Highlight the file you want to retrieve.
3. Right-click and choose *Restore*, or move your mouse to the *Restore item* option in the left sidebar.

It is very important to,
every so often,

Empty the Recycle Bin!

When you are sure that there is NOTHING in the Recycle bin that you want to retrieve, empty it.

Here's how:

1. Open the Recycle bin
2. In the left sidebar, click on *Empty Recycle Bin.*

Once you have emptied the Recycle Bin, you will NOT be able to retrieve these items ever again!

Windows Explorer

There is a very cool program in your computer called ***Windows Explorer***.

With *Windows Explorer* you can see all the files and all the folders in your computer, making it **very easy** to move or delete any file. Really!

Here's where to find Windows Explorer:

Start > Programs > Windows Explorer

or

Start > Programs > Accessories > Windows Explorer

Notice that the screen is divided into two sides.

- The left side shows *all your FOLDERS.*
- **When a folder on the left is highlighted**, the right side window will show all of its contents. You might see folders within folders and many files.

This makes more sense if you have the program open!

Windows Explorer

On this side of Windows Explorer you will see the different Drives in your computer and Folders that contain programs and documents.	Click on a folder on the left side and you will see its contents on this side! Here, "My Documents" is open, showing the folders and files inside. Highlight a file on this side to delete or move it.
Desktop	Basketball
My Computer	My Book Drafts
My Network	School Stuff
My Documents	First Save
Recyle Bin	Second Save

To delete a file or folder in Windows Explorer:

Be careful not to delete files or folders that are imperative to your computer!

***Only delete your own stuff. ***

1. **Left-click** a file or folder to highlight it, **right-click**, choose Delete, **left-click**

Select
Open
Print
Copy
Create Shortcut
Delete
Rename
Properties

2. It will ask you if you really want to delete this. Click *Yes* if you do, *No* if you don't!

To Move a file or folder it's just

CLICK & DRAG!

This is too easy to be true!

1. Click through the folders shown in **the left side window** until the file you are looking for shows up in the **right-side window**.
2. Now, back **in the left-side window**, scroll around, up and down, until you can **see the folder** you want the file moved to. As long as you don't click on anything, the right side window will stay the same!
3. In the right-side window, left-click on the file you're going to move to highlight it. CAPTURE it, by **holding the left-click down. Now, with the mouse still clicked down, simply drag the file** directly to, and over, the name of the folder you want to deposit it in.
4. Release your mouse. - That's it! - Moved!

If only moving day was this easy....

Games

This doesn't have anything to do with customizing your computer, but it's important to remember to play.

Don't forget, you can just have fun with your computer too! Lucky for us, most computers come with a couple of games pre-installed!

Some of the more common games are Solitaire, Hearts, Pinball and Backgammon. In Windows XP and Vista, you'll also see internet games! These are all really fun to play and you don't have to give any personal information to play them. Just click to open and follow a couple of prompts. *It's that easy!*

Here's how to find them:

1. Start > Programs > Games
2. Click on the game you want to open.
3. PLAY!

Revving up Word!

Revving up Word!

Customize Word

Changing Default Font
If you don't like the size or style of typing that appears every time you open *Word* you can **change the default setting** to almost anything you like!

Larger Smaller *Italic* Comic Papyrus Gigi
Almost anything you like!

Here's how:

1. Open *Word*

2. To open the *font window* click: Format > Font

3. Within the font window you can choose the size and style of fonts. Scroll through and highlight the style of font you want.

4. Once you have chosen what you want, you **MUST click on the *Default* button** at the bottom left-hand corner of this window.

   If you click on OK, you will only affect your current document.

5. *Word* will ask if you are sure you want to do this, as it will **"change the global template."** It might also refer to the "normal" template.

   Sounds very ominous... Say YES!

Now every time you open Word, it will look the way YOU want it to look. That's very nice indeed.

Customize Word

Spelling and Grammar
This tool checks spelling and grammar automatically as you type.

To point out mistakes, *Word* will place a wavy red underline to show possible spelling mistakes, and a wavy green underline, to show possible errors with your grammar. **These lines do not show up when you print the document.**

Did you notice I said "possible" mistakes or errors?

Spell check is only as smart as its programmer; they're pretty smart, but... For instance; *Word* will think an unknown name is a mistake, like *Latremouille!*

When you have typed a word that you know is right, but your computer doesn't recognize, you have to *Add* the new word to its dictionary.

Next, how to customize Spelling and Grammar!

Customize Word

Customizing Spelling and Grammar and more!

Here's how:

1. Tools > Options > Spelling & Grammar
2. Look at the options, tick them on or off

This is the place for a bunch of options, such as:

- Your User Information (your name and such)
- How Word treats Edits
- Default File Locations
- Setting document Security, like passwords
- Default Print settings
- General options, like sounds or animations
- Save options, like how often to auto-save
- Track Changes, how this looks and works
- Compatibility, to other version Word documents
- View, for default view settings

Click into all of your options and see what's what.

TIP

If you are not sure about a change, write down what the current setting is, BEFORE you tick, or un-tick things!

Tailored to a perfect fit!

Keyboard Shortcuts

Letting go of the keyboard, grabbing the mouse, then back to your keyboard, doesn't sound like much. But if you have to do it repeatedly, it might drive you nuts! That's why it is nice to know at least a couple of shortcuts!

Almost anything you do with your mouse, you can do with a keyboard shortcut. Using shortcuts can make a world of difference, if you use a command quite often.

Key points about using keyboard shortcuts:

- The combination of keys must be *pressed together* to make the shortcut active.
- You can use a shortcut on highlighted material
- You can use a shortcut to change the font of what you will type next.
- Use it once to turn it on, use it again to turn it off!

Keyboard Shortcuts

Here is a list of common shortcuts

- CTRL + B   Type **bold**
- CTRL + I   Type *italic*
- CTRL + U   Type <u>underlined</u>
- CTRL + C   Copy the selected text or graphic
- CTRL + X   Erase the selected text or graphic
- CTRL + P   Print
- CTRL + Z   Undo the last action
- CTRL + Y   Redo the last action
- CTRL + Shift + **<**   Decrease font size
- CTRL + Shift + **>**   Increase font size

Sometimes a shortcut IS a good idea!

Bright Ideas

◇

◇

◇

◇

◇

◇

◇

◇

Letters & Mailings

I'm always learning new things in Word! And, the day I finally felt brave enough to try Letters and Mailings was a good one! Find it under Tools.

Tools > Letters & Mailings

Letters & Mailings	**Mail Merge Wizard...**
	Show Mail Merge Toolbar
	Envelopes & Labels
	Letter Wizard

Mail Merge Wizard...

Mail Merge is what companies use to personalize form letters. You can use the Mail Merge Wizard to help you create personalized:

- Letters
- Email Messages
- Envelopes
- Labels
- Directories

"Merge" - a great word that explains alot! This function looks for information, maybe a list of addresses or names, from somewhere else in your computer; then, puts that information **into** another document! *How cool is that!*

Letters & Mailings

How does Mail Merge work?
Simply put, the programs talk to each other. As always, it's easier to understand after you do it, so...

- Open Outlook's Address Book
  (If you don't use Outlook, open what you've got.)

- The address book will be your **Data-Source**.

- Now, open up someone in your address book, a *Contact*, to see the information (properties) you have for them.

- The spaces where you see their name, address, phone number, email address, etc., are called fields.

- **<<Fields>>** This term is important to know.

- The more complete the fields are in your data-source, the more mail merge can do for you.

When you use Mail Merge, you insert various <<fields>> into your letter or document. For example; if I want to send the same letter to 10 people from my address book, parts of the letter on my computer screen might look like this:

Dear <<First>>,

Look at your contact's information. See the field where it says "First"? Word looks for the specified field, then inserts that information into your letter!

Let's Practice a Mail Merge!

Open a new document in Word, then
Tools > Letters & Mailings> Mail Merge Wizard

1. In the Mail Merge Wizard sidebar that opens, near the top, choose "**Letters**"
- Then, near the bottom of the sidebar, click on "**Next**"

2. Back to the top of the sidebar, under *Select starting document,* choose "**Use the current document"**. Yes, this document is blank, Don't worry, you will start typing a letter soon!
- Again, at the bottom of the sidebar, click on "**Next**"

3. Back to the top of the sidebar, under *Select recipients* choose: "**Select from Outlook contacts"**
- Click on "**Choose contact folder**"
- A small window pops up, with "Contacts" highlighted, click OK. - *So far, so good?*

4. The *Mail Merge Recipients* window opens.
- All of your contacts are listed here.
- Notice the headings: Last, First, Title, etc....
- These headings can correspond to "Fields" in your letter.
- If you don't want to send a letter to everyone in your address book, click on the "Clear All" button. That un-ticks the list. Now you can click on, to include, only the recipients you want. Click OK.

5. Again at the bottom of the sidebar, click on "**Next**".

   Now you can start to type your letter!

6. Back to the top of the sidebar, under "Write your letter" you will see a few options: Address Block, Greeting Line, Electronic Postage and More Options.
   - Address Block: click on this, to pick how you want the address to look on your letter. It will be inserted where ever your cursor is.

     Try it out. Practice. See what I mean.
   - Greeting Line: this helps format the opening.

   Here is an example of what it will look like:

> ««**AddressBlock**»»
>
> Dear «**Title**»«**Last**»,
>
> I am getting the hang of Mail Merge.
>
> I can insert your email address like this «**Email_Address**» and even your company name like this «**Company**».
>
> Sincerely,

When I finish the merge, the email and company name (found in your address book) will automatically go into the fields between the brackets. The brackets don't show up on the final copy.

7. I want a formal letter, so after I type Dear, I have to add a field for my title (Ms.) and another for my last name.

   Click on "**More items**"

In *More items*, you will see all the different types of fields you have to work with. **Choose an item**, click on "**Insert**" to send the field to your letter, then click on "**Close**". Back on your letter, you'll see the field.

9. Again at the bottom of the sidebar, click on "**Next**" Preview your letters!

   Your letter will be transformed before your eyes!

> **Ms. Louise Latremouille**
> **P.O. Box 3100**
> **Tantallon,, Nova Scotia B3Z 4G9**
>
> Dear **Ms.Latremouille**,
>
> I am getting the hang of Mail Merge.
>
> I can insert your email address like this **louise@myparentsfirst.com** and even your company name like this **KLMK Enterprises**.
>
> Sincerely,

10. At the bottom of the sidebar, click on "**Complete the Merge**".

Letters & Mailings

Need a drum roll here!

Print your letter and you're done!

I'll bet, even though it took 3 pages and 10 steps it was way easier than you thought!

Notes____

Letters & Mailings

Letters & Mailings	Mail Merge Wizard...
	Show Mail Merge Toolbar
	Envelopes & Labels
	Letter Wizard

Show Mail Merge Toolbar

Click this option to have the Merge tools as part of the menu bar. This is handy if you are creating lots of form letters.

Envelopes and Labels

Tools > Letters & Mailings > Envelopes & Labels

If you want to print out envelopes or labels you need to tell your printer what size of envelope or label you are using. This tool lets you do just that.

Click on Envelopes & Labels and follow along on the next page to discover how easy it can be!

TIP!

Buying a label that is listed in the label "options" (and most are!) makes it easy for your printer to print the information in just the right spot!

Letters & Mailings

Printing on Envelopes

Open the "**Envelope**" tab:

- Type in the delivery and return addresses in the windows provided.
- Click on "Options" and tell it what size of envelope you have.
- Put your envelope in the printer and then OK!

Printing on Labels

This is a great feature if you would like to print out labels for canning, wine, CDs, addresses, whatever!

Open the **"Labels"** tab:

- Click on "Options" and choose the type and size of label you're using.

Now you have a couple of choices. You can either type the information in the space provided, and click on Print, and you're done. Or..... Do what I like to do!

- Click on the "New Document" button to see a shadow version of the label you selected, open up as a new document.
- Type what you want. It's easy to fancy up the labels here, with different fonts or even add images if you want!
- Click on Print and you're done! *Ta Da!*

Letters & Mailings

The Letter Wizard
The Letter Wizard helps you *format and address* letters.

When you open the Letter Wizard, you will see different styles and formats to choose from; from formal to friendly. It can help you choose your salutations, closings -- and even insert addresses!

Using the *Letter Wizard* to address letters is easy, if you use the Address Book in *Outlook*. If you don't use Outlook, you will have to tell the Wizard where to find your Data Source.

When you open the Letter Wizard, notice the four tabs on the window that pops up:

Letter Format	Recipient Info	Other Elements	Sender Info

Click on each tab and have a look around. It is very self explanatory and easy to use.

Under *Recipient Info* you will see a little icon that looks like an address book. Click on the icon to zoom to your own address book and choose who you are writing to.

That's it!

If only I knew before, how easy this was!

The only thing left to do is write the letter!

Bright Ideas

◇

◇

◇

◇

◇

◇

◇

◇

Microsoft Excel in a Nutshell!

Microsoft Excel in a Nutshell!

Excel

Microsoft Excel was designed with accounting in mind, and if it was just limited to that, you could think of it simply as an accountant's ledger. But it does so much more! On the next few pages I'll show you the basics.

Excel is a *spreadsheet* program. All that means is, it is set up like a big table! Open Microsoft Excel® and you will see what I mean.

Start > Programs > Microsoft Excel

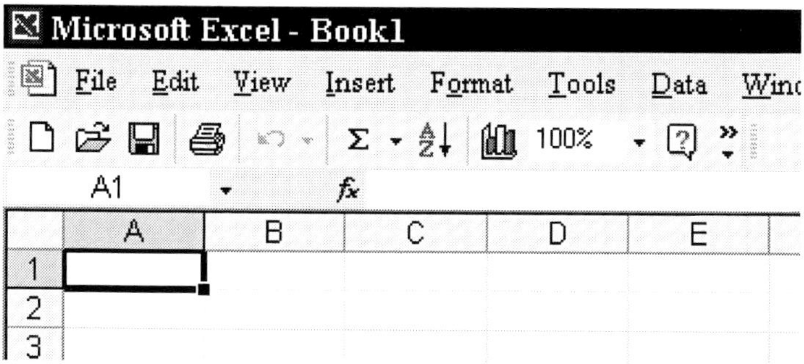

Excel is all about the grid. Each cell (box) in the grid has a name assigned to it because of its location.

For instance, the cell in the very top left is named A1. The cell beside it, B1. Can you figure out where cell C2 is? Or where cell E3 is?

Notice that cell A1 has a box around it. That means that the cursor is active in that cell.

Also, see where it says A1 in the space just above the cell? *That space ALWAYS* reflects what ever cell your cursor is active in.

Excel

Here I have typed "LABEL" in cell B2. Notice that "LABEL" is also shown in a window above. This is the function window.

"*fx*"
f : for function
x : for whatever number you use

When your cursor is active in a cell, whatever is in that cell will automatically show up in this space.

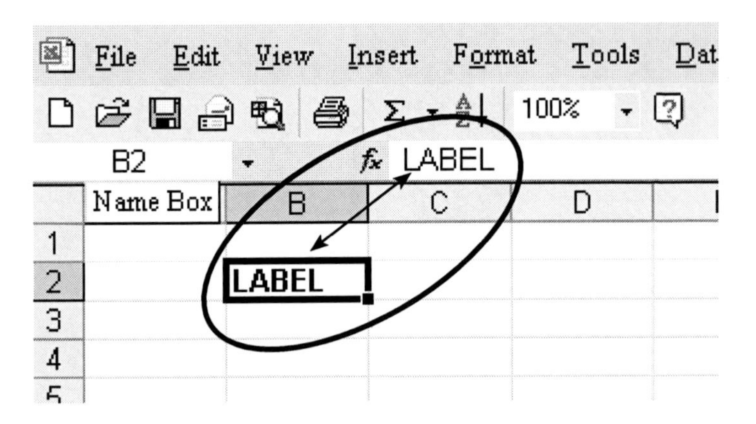

Whenever you learn a new program, you have to learn some new lingo that goes with it. Excel is no different. Here are a couple of new terms:

Label: the name of a cell with typing in it
Value: the name of a cell with numbers in it

Knowing these two terms will make it easier to understand instructions in Help, or when you get an *error* window pop up.

Excel

When you use Excel for any math function, you have to create a formula to tell it what you want.

For example, you would enter the formula, =1+2 in a cell for the equation, 1+2=3. Excel does the math for you. You can do any sort of math with Excel.

Here is a list of math symbols found on your keyboard. You can use them to create formulas.

=	Equal
>	Greater than
<	Less than
>=	Greater than or equal to
<=	Less than or equal to
<>	Not equal to
+	Add
-	Subtract
/	Divide
*	Multiply
%	Percent
^	Exponent (to the power of)

These symbols are also called Arithmetic Operators, or in Excel, just Operator.

You use a math operation (2+2) to solve a math problem; hence, the operator in the middle of the numbers. You can have lots of operators in a problem. Remember problems like this? 6 + 3 - 4 x 8 =

Excel

Creating a Formula

If you want Excel to do math for you, you have to let it know your intentions. You have to let it know you are seeking an answer to your question before you ask it. *How Confucius!* So, the equal sign = is placed in front of the question.

While in regular math you write the *equation* 1+2=
in Excel you would type a *formula* =1+2

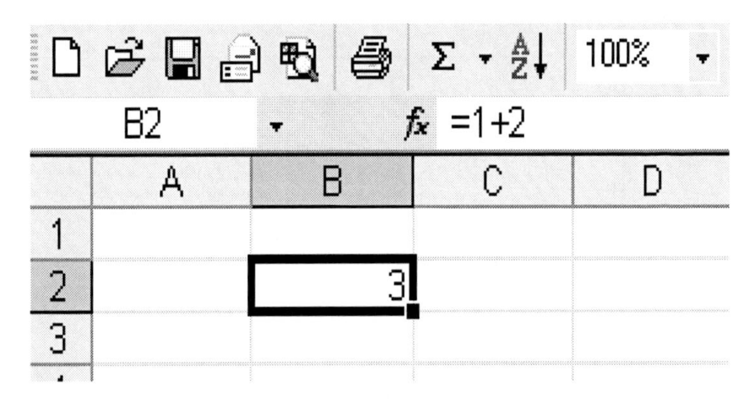

Here I typed =1+2 in Cell B2, then hit the Enter Key. **Notice that only the answer - or result - is in the cell. What I typed, is showing in the *fx* (function) window.**

Results of formulas always show in the cells where they are typed. You have to remember to look back into the formula window to see what you actually typed in there!

Easy, peasy, nice and Weezie!

EDITING A CELL
Once something is entered into a cell, the only way back into it, is by typing directly into the formula window!

So, using the example on the last page; if I wanted to change =1+2 into =2+2, I would click the cursor into the formula window and make the change.

Practice! Try it out!
The best way to learn, is to do.
So please, practice as we go along.
You don't learn to ride a bike
by reading about it!

Adding different cells together

Being able to use information in a variety of cells in whatever combination you need is one of the things that make a spreadsheet program so versatile.

Here I asked
cell C3 to figure the sum of cell A1 and B2

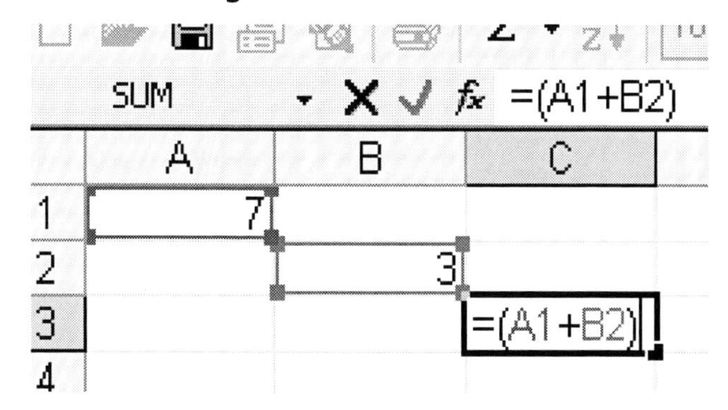

Excel

As soon as you hit Enter, the answer shows up in the cell. Notice that the formula (what I typed into the cell) is in the *fx* window.

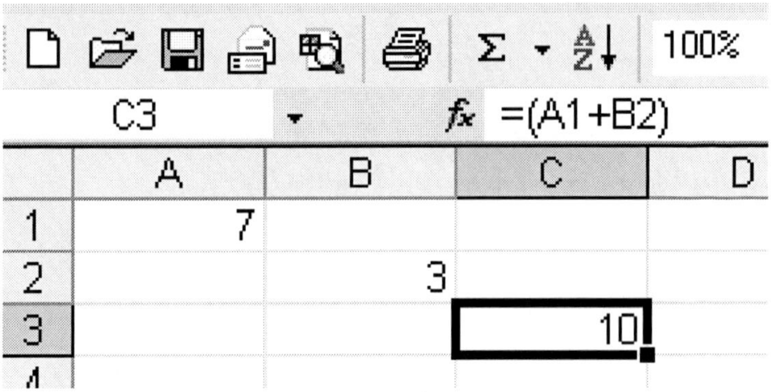

100%

C3 *fx* =(A1+B2)

	A	B	C	D
1	7			
2		3		
3			10	

In this next example I changed the formula to
subtract the cells, by changing the *operator*.
In the *fx* window, just change the **+** sign to a **-** sign!
You can change the operator anytime you like.
\+ - / *

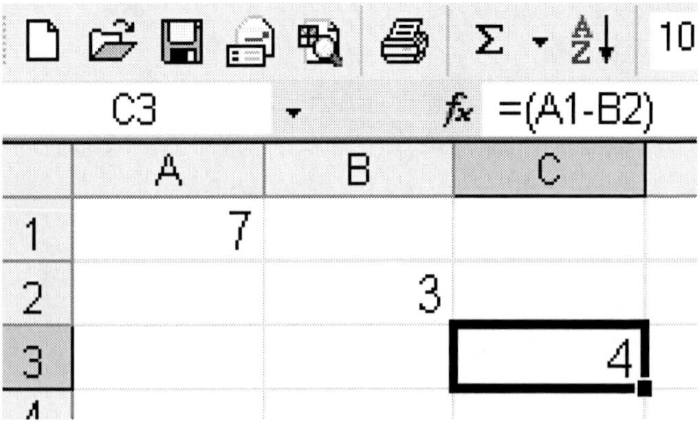

10

C3 *fx* =(A1-B2)

	A	B	C
1	7		
2		3	
3			4

Try it out! Practice! Follow along!

Adding entire columns or rows - and a shortcut!

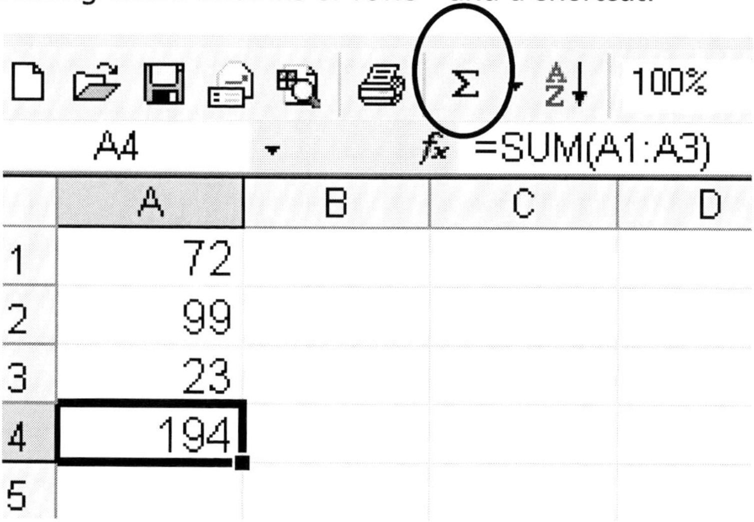

The AutoSum tool

AutoSum is a great tool, if you have a column, or a row of numbers that you want to total up.
Here's how to use it:

1. Click your cursor into the empty cell at the bottom of your column of numbers
2. Move your mouse and click on the AutoSum

Notice that the *fx* window shows the formula. When you want to add many cells together you can use a colon (:). In this example, I've asked Excel to add cell A1 all the way through to A3, inclusively.

If you want to AutoSum a row, you have to highlight it first. Then with the cursor in the empty cell at the end, click on AutoSum.

Sorting Data - Organizing a List.

Here's how:

1. Highlight the material you want to sort.
2. Move your mouse to Data > Sort
3. A window, similar to this one, will pop up.

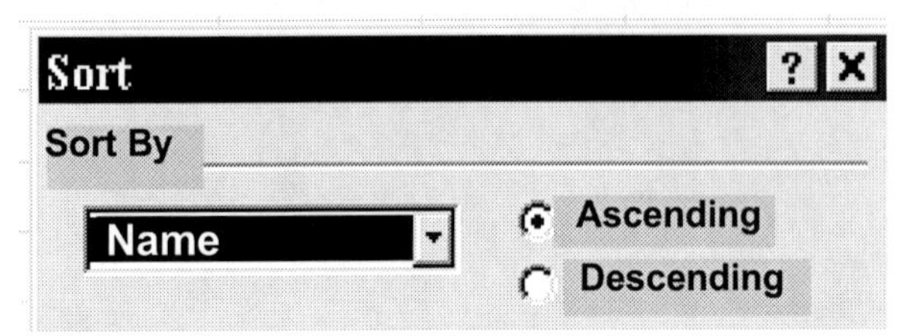

4. Choose how you want to sort it. On the example, I made headings: Name & Age. You can sort by the columns too: A, B, C, etc.

Have a look at the lists on the opposite page. One is sorted alphabetically by name, the other numerically.

> View > Toolbar > Standard
> Make sure that "Standard" is ticked off, as one of your toolbar options! That will give you all the icons beside the 100% window shown in the examples. Very handy!

That's Excel basics in a nutshell. Play around with it and remember to save your work as you go along. That way if you mess up... it won't matter too much!

Excel

Sorted by Name

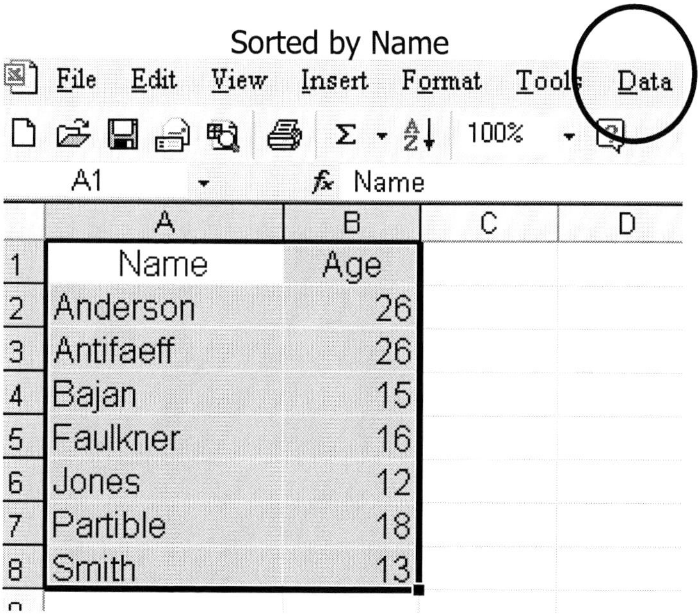

File Edit View Insert Format Tools Data

A1 fx Name

	A	B	C	D
1	Name	Age		
2	Anderson	26		
3	Antifaeff	26		
4	Bajan	15		
5	Faulkner	16		
6	Jones	12		
7	Partible	18		
8	Smith	13		

Sorted by Age

File Edit View Insert Format Tools Data

A1 fx Name

	A	B	C	D
1	Name	Age		
2	Jones	12		
3	Smith	13		
4	Bajan	15		
5	Faulkner	16		
6	Partible	18		
7	Anderson	26		
8	Antifaeff	26		

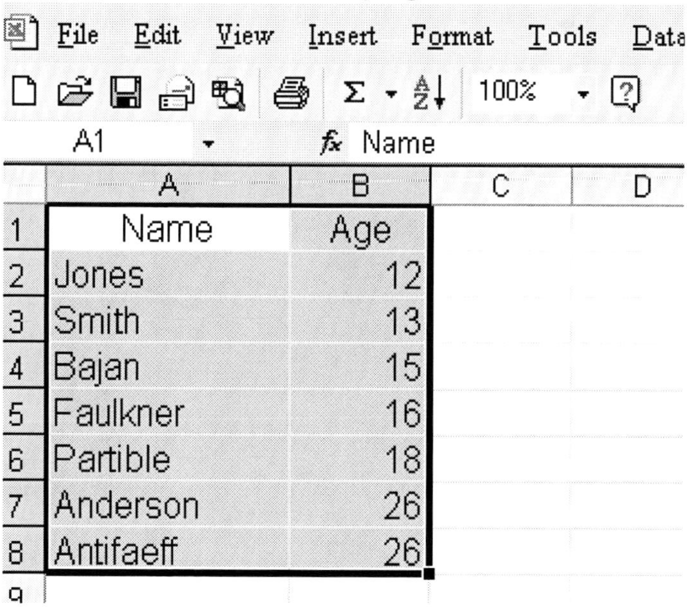

Bright Ideas

-
-
-
-
-
-
-

Internet Savvy

Internet Savvy

Internet Savvy

Seems the more I use my computer, the more I expect it to do. I feel the same about the internet. I want to use the internet, not have the internet use me.

Although using the internet is easy, not letting it waste your time is another matter. Most importantly, you need to know how to keep you and yours safe!

Savvy? - Read the next little bit and you will be!

Internet Security

Once upon a time, not too very long ago, speaking privately on a telephone required a leap-of-faith.

Learning to trust the internet can require this same leap.

When you are finally ready to purchase goods on-line with a credit card, or do your banking on-line, you should do it with more than faith.

You should enter this world
with a bit of know-how.

So, on the next few pages we'll learn some **essentials about internet security.**

Internet Security

There are four important aspects when considering internet security:

1. **A Firewall** – Denies access to your computer from outside, uninvited guests.

2. **Anti Virus Protection** – Your best protection against known computer viruses from entering or leaving your computer.

3. **Automatic Updates** – Continually updates Firewall and Anti Virus programs.

4. **128 bit Encryption** – Encryption is a way of scrambling information via two computers. With encryption, digital keys are needed to send and receive the information. With *128 bit encryption* the digital key combinations are in the millions.

Internet Security

If understanding internet security is a little fuzzy, this might help clear things up. Consider the three little pigs...

The first little piggy: straw house, flimsy windows and door - No Security!

The second little piggy: a solid wooden house with a couple of locks, but the big bad wolf wouldn't have too much trouble breaking in. Only security here - a Firewall.

The third little piggy went all out. Strong brick house, reinforced windows, dead bolts on the doors and an alarm system. Safe and secure. He's got current Firewall and Anti-virus programs and does automatic updates.

***You* should live in house #3!**

When you bank on-line or use your credit card, you need even more security! And, the security has to come from *who and where* you are dealing with.

Internet Security

Your bank knows about the big bad wolf and they won't let him in!

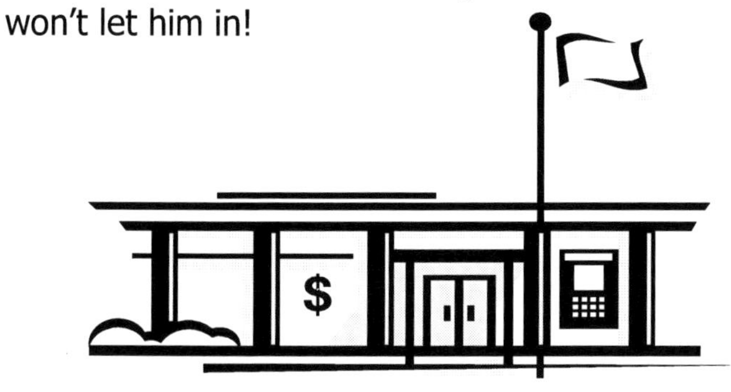

...locked steel doors, barred windows, guards...

Their computer, along with frequently updated Firewall and Anti Virus protection, uses secure 128 bit encryption for communicating.

Any time you give your credit card or other personal information, make sure the company you're giving it to uses encryption, 128 bit or higher!

There are two ways your computer will show you that you are using a secure site:

- The web address will start with **https://** (see the "s" at the end)

- A security icon will show in the task bar.

When you use the internet for banking or purchasing, ensure that you are **communicating privately and securely.**

Filtering Internet Content

Internet Content

You may want to set *parental controls* on your computer to filter content such as offensive language, nudity, gambling, drugs etc. from being downloaded.

You can set some boundaries using tools found in the *Control Panel.*

Control Panel> Internet Options> Content tab> Content Advisor > tick on, Enable

You will create a password for yourself and then be able to choose the level of "content" you will allow. You also have an option here to completely block certain sites, or to only allow certain sites.

For more information on content rating I suggest you have a look at the Internet Content Rating Association's web site: **http://www.icra.org.**

Downloading

You can download almost anything from the internet into your computer. Besides considering that you are downloading something clean of viruses, you should also consider...

Download Speeds
The speed of *your own* computer and internet connection is not the only consideration, for sending and receiving data.

Data can only be received as *fast as the senders computer can send it,* or as fast as a modem can receive it.

If you send a large file to a very slow modem, it could really bung up the receiver's system.

Keep your friends happy 

Don't send large files to them unless you know their system is fast enough to receive it!

Pictures can be large files.

Downloading

Downloading
All types of files and programs can be downloaded. A program will contain different types of files within it.

You don't really have to know about file extensions, but it's one of those things that make understanding your computer's quirks a little easier.

All *programs* are *applications*. *Application files* all end with *.exe*

Here are a few common file extensions.

.exe	=	executable application
.jpg	=	image file
.gif	=	image file
.bmp	=	image file
.doc	=	document file
.dll	=	needed to help applications
.txt	=	text file
.mp3	=	audio file
.wmv	=	windows video file
.pdf	=	Adobe portable document
.ppt	=	Power Point file
.xls	=	Excel file
.mpeg	=	movie file
.wpd	=	Word Perfect document
.zip	=	zipped file (super condensed)

Downloading

Downloading files or programs
You will find lots of reasons to download files or programs from the internet. You might even need to download a new program to view or open a file.

Automatic downloads
Many programs are pre-programmed to update themselves. Firewalls and internet security programs darn well better be set to update themselves!

Programs do this by logging onto their maker's server, and automatically downloading any updates needed.

Internet cache
If you are browsing the internet, your computer will be downloading little bits and pieces all the time without you even knowing it.

These little bits are called *cache or Temporary Internet Files*. They help web sites load faster the next time you visit them.

You can tell *Netscape* or *Explorer* how much internet cache to accept. And, like your trash you should empty your cache once in a while.

In fact, you should absolutely clear your cache, or temporary internet files, after doing any on-line banking!

Downloading

Before you download:

- Know *where in your computer* you are going to put your download. I have a folder in my computer called, Downloaded Programs.
- Make sure you have an anti-virus program. After I download a program or file, **before** I open it, I ask my anti-virus program to check it over!
- Know what type of system you are running. Vista, Windows XP, Windows 98, etc. Will what you are downloading, work on your system?
- Consider how big the program or file is. How long will it take to download through your modem.
- Is it a free download or do you have to pay for it? Will you need your credit card handy?

After you click on a *download now* button, a window will pop up that:

- Will show the name of the file. This often is not the name of the program, write down the name of the file or change it to something more appropriate - don't forget about its file extension!
- Will suggest what folder it wants to download into. *Note that too! Nothing worse than looking for something you just downloaded!*

Bright Ideas

◊

◊

◊

◊

◊

◊

◊

◊

◊

◊

Banking On-line

Every day, more and more people are enjoying the benefits and ease of banking on-line. Why? No long line-ups, instant access to your accounts, easy bill payments, the ability to transfer funds between accounts, even view investment portfolios!

The best part of banking on-line is that it's simple!

Banks want their customers to use this service, so they have made it user friendly. Some banks will charge for the service but many banks offer on-line service for free (it's included with your regular banking fees, whether you use it or not!).

I talked about secure sites earlier and 128 bit encryption. Any reputable bank uses 128 bit encryption for internet services.

Like most other things in life, banking involves trust. Trust your bank, but ask about encryption, account security and fraud protection before you sign on!

Banking On-line

Different banks will have different looking web pages, but for the most part they offer the same services.

Here are a few of the most common services:

- Real-time access to account balances
- Bill paying
- Transfering funds between accounts

Here's what you need to sign up:

1. You must have a current account at the bank.

2. You will need your Social Security Number (US) or Social Insurance Number (Can).

3. You will need a valid e-mail address.

4. Your browser must be able to handle 128 bit encryption.

5. You will be asked to create your own personal secret User ID and a secret Passcode.

Once you have these basic things, signing up is easy, just follow your banks on-line instructions!

Banking On-line

You're all signed up, now what?

You will find that banking on-line is pretty straight forward.

Go to your bank's web site and "sign-on" in the space provided. You might need your user name and a password, or you might just sign in with your account number and a password.

Once you are signed in, you can:

- See your account balances
- View all of your accounts at one time
- Do a variety of banking chores

When you bank on-line, you hold all the strings when it comes to paying a bill. You say who, how much and when. If you want, you can even set up automatic bill payments!

There is a catch! The billing company must have a contract with the bank, allowing their bills to be paid there.

Most utility and large companies are registered with banks, but there is a good chance Joe's Plumbing isn't set up for on-line bill payment!

Banking On-line

Pay your bills on-line! Here's how:

1. Log onto your bank
2. Look for the link that reflects "Pay Bills On-line"

You will have to "add" your billing companies to your on-line banking. Look for a button (link) on this page that reflects that.

3. Search for your billing company and place them on your list

Once you've added a company, you can pay them!

4. Pick the company you want to pay
5. Select which account you want to pay them from
6. Enter the amount of the bill
7. Confirm the information
8. Click on Pay
9. The bank should give you a confirmation or *reference number*, write this number down on your bill. You can even print this page for peace of mind.

That's it! I said once before that, the real reason so many people are using computers is because it is so easy. You are starting to believe me; right?

ONE MORE IMPORTANT THING

–

SECURITY!

Banking On-line

Security

Have you heard of the term "hacker"?

A hacker is someone who sneaks into a computer for information. An information robber.

If you are ever a victim to this crime, you certainly don't want any banking information available in your system. No worries, there's a fast and easy way to protect yourself.

Whenever you are on a web site, your computer constantly downloads little bits of information. This stored information is called *cache or temporary internet files.*

If you don't leave money laying around on the kitchen table, why would you leave it laying around in your computer? The cache in your computer can be just like "money on the table" for a hacker.

This is why it is important that,

after you log-off from your bank,

you must clear your computer's

Temporary Internet Files (or Cache).

Heres how to clear cache in Netscape, 6.2 or higher

1. Click on *Edit* > Preferences
2. In the left side-bar, click on *Advanced.*
3. Select *Cache.* Click on the *Clear Disk Cache* button, then *OK.*
4. Click on the *Clear Memory Cache* button, *OK.*

Heres how to clear Temporary Internet Files in Internet Explorer 5.0 or higher

1. Click on *Tools* > *Internet Options*
2. Under the *General* tab: by *Temporary Internet files* or *Browsing history*, click on *Delete or Delete Files.* A new little window might open here.
3. Click on the *Delete Files* button.
4. If you see the option to *Delete all offline content,* tick it, then click OK.
5. When clearing is complete, click OK.

You might see options to delete Cookies, History, Form Data, Passwords or maybe an option to Delete All. Deleting any or all of this information won't harm your computer. In fact, every once in a while you should clear out these temporary files!

Bright Ideas

- ◇
- ◇
- ◇
- ◇
- ◇
- ◇
- ◇
- ◇
- ◇
- ◇

EBAY! The world's largest virtual garage sale!

OK, so maybe it's an Auction House, but I've always thought of it as a garage sale – something I'm more familiar and comfortable with!

Like any garage sale, you can stroll in and browse at all the items. If you see something you like you can look at it a little closer. If you REALLY like it, you're going to have to introduce yourself to the seller, and make them an offer!

eBay is not complicated and is set up to give the buyer and the seller confidence. It is world-wide and browsing is free. Here are some of the basics we'll cover on the next few pages:

- Where is www.ebay.com
- How to search for items by category
- To buy or sell, you must register with eBay
- How to buy
- How to sell

eBay

Where is eBay?

www.ebay.com is based in America, www.ebay.ca is based in Canada, www.ebay.de is based in Germany.

eBay is all over the world. Each country following standard practices that has made eBay what it is, a service built on trust. They are connected, in that once you are registered with one, you are registered with all.

You can find almost anything on eBay. Search and you will find what you are looking for.

In fact, it's easy to find *too much* of what you're looking for. So, here are some tips to make searching easier.

How to search eBay:

It's easier to learn by doing, so...
let's practice!
Please log-on to www.ebay.com

- **Be precise**. Let's search for a TV. Today, as I write this, I searched for "television" eBay found 762 items. Then I tried "tv" and it came up with 11,266 items. Yikes!

- Notice in the left side-bar, "Matching Categories". This can help you narrow down your search. I chose to narrow my search down to "televisions", now it shows 5,477 items. Still, Yikes!

- The left side-bar now offers ways to narrow down your search even more. Tell it more of what you are looking for, like the style or type of tv, the screen size, the brand.

- Choose a few options and click on "Show Items" I chose: Portable, LCD monitor technology, and 5 – 9 inches for the screen size. **7 Items!** – *Now this is a number I can deal with!*

eBay

Items listed on eBay are updated every hour. The lists are ordered by date, the most current being first. *Regular eBay users can qualify to purchase an option that will promote their item to the top of the list.* What you find one hour, you might not find the next.

Items are listed, described briefly, and categorized under these headings:

- **Compare:** You can tick multiple items you are interested in and then eBay will create a new list showing only these items.

- **Item Title:** What the seller has chosen to name their item.

- **Price:** Shows the asking price, minimum price acceptable price (reserve) and current bid. "Buy it now" means that if you pay that price, it's yours!

- **Bids:** The number of bids that have been placed on an item.

- **Time Left:** Shows the length of time the listing has left.

If you see something you're interested in, click on its picture or title, to see more information about it.

Tip:
Use the "back" button on your browser to return to previous screens when surfing eBay

If you click on an item for more information, this is what you will see:

- **Starting Bid:** The starting price the seller placed on it.
- **Current Bid:** Highest and most current bid on the item.
- **Time Left:** Remaining time for the listing.
- **Start Time:** The day and time the item was listed.
- **History:** Lists the bidding history, which includes who and how much was bid. This is a good place to find honest comments from buyers and sellers.
- **Item Location:** Where the item is physically located.
- **Featured Plus Listing**: If applicable, this will show up here – Featured Plus is the selling option that will bring an item *near* the top of the list.
- **Ships To:** Where the seller is willing to ship to.
- **Shipping Costs:** How much, by what method the item would be shipped.

Scroll down the page to find a *description* of the item. This is what the seller has written up to promote it.

The description area is a good place to discover if you are buying from a business or from a regular Joe down the street.

Many businesses sell on eBay.

eBay

See something you absolutely must have? To buy or sell anything, you must register with eBay. Don't worry, it's quick and easy to register.

HERE'S HOW TO REGISTER

Click on the "Register Now" button to be led through a **3-step process**, asking you to:

1. Enter your personal information.
2. Create your User ID and a Password.
3. Confirm your registration.

Step 1, Personal Information

- You will be asked for personal information such as your name, phone number, address and e-mail.

- Credit card information is not always required, but, in the event that you are asked:

Quoted from eBay.com

"You will be asked to place your credit card on file if the e-mail address you entered cannot be used to verify your identity. This usually happens for one of the following reasons:

1. *The e-mail address came from a free, web-based account such as a Yahoo or Hotmail account.*
2. *The e-mail address is invalid - that is, attempts to send a message to this e-mail address repeatedly "bounce" back to eBay."*

End quote.

- You will be shown the User Agreement and Privacy Policy and have the choice to *accept* or *not accept* it. **If you *do not accept*** the agreement – you're done here!

If you accept the agreement, you go to Step 2!

eBay

Step 2, User ID and Password

- Create a *User ID,* it does not have to be your real name. Have a look on eBay and see what kind of user ID's people have. Choose a name that you will remember and write it down somewhere!
- Choose a password.
- You will also be asked to create a *question and answer* that can be used to identify you if ever you forget your ID and password.
- If you forget your User ID or Password, eBay will ask you your secret question, among other personal stats, and then e-mail the information to you.

Step 3, Confirming your Registration

- Next, you will receive an e-mail from eBay that asks you to confirm your registration.
- Make note of the *confirmation code number.*
- Click on the *Complete eBay Registration* button.

That's it.
You're ready to
shop on eBay!

Bidding & Buying!

How do you pay?

The Seller determines how they want to be paid. Visa, BidPay, MasterCard, American Express, PayPal, personal cheque, money order and more; even cash! **I don't advise sending cash through the mail.**

Before you bid on an item, know these things:

- Be sure that you really want to buy an item before you place a bid.
- Look for a reserve price. Sellers often set a minimum price (the reserve) that must be met.
- Look for the shipping fees. Some items are priced very low and make up for it with high shipping fees! *Tricky people...*
- See what form of payment the seller will accept, make sure you're comfortable with that method.
- Where is the item coming from, and will the seller ship to your area?
- Notice if insurance is offered.
- Notice if they are new to eBay or have a history. If the seller has a history, check it out!

eBay

Bidding & Buying

You have found the item that you *just must have!* Yippee!

Before you go ahead and bid on it, decide what your TOP dollar is going to be. You might be surprised just how much like an actual auction eBay can feel like. It is very easy to get carried away! It's fun bidding on items, and even when the amounts climb at only 25 cents a bid, the quarters can add up pretty fast!

Here's how to bid:

1. Click on the item's title to open up the information page. Click on *Place Bid*.

2. Enter the dollar amount you want to bid in the *bidding window* that pops up.
   - ◊ Notice the statement "You are agreeing to a contract..." by confirming the bid.

3. Click on *Confirm Bid* to see if your bid will be accepted. It might be good, or it might be beat out by a previous bid. Decide if you want to bid again.

4. If you have a winning bid, you will receive an e-mail confirming your bid. If you get outbid, you'll get another e-mail letting you know!

You can also do **proxy bidding** – tell eBay your maximum bid, and they will bid for you.

Bidding & Buying

You're the highest bidder! You won! It's yours!
Now you have to pay for it.

$$$

Here's how to pay:

1. eBay will send you an e-mail: "Congratulations – You Are The Winning Buyer!"

2. Check for payment options and details within the e-mail.

3. If the shipping amount is yet to be determined, the Seller might e-mail and ask for a destination and then let you know.

4. The Seller will send the item once your payment is processed. If you pay with a check, they will probably wait till it clears.

Remember. Use the "back" button on your browser to return to previous pages.

eBay

Selling

Don't be fooled into thinking you can make millions selling items on eBay. Just like you, people are shopping on eBay looking for deals. Whatever you sell, unless it was a gift, you paid something for it.

You'll find lots of selling tips on eBay. Some tips are just good suggestions. Other tips can be an added expense, such as *Featured Plus* that puts your item near the top of the heap.

When you list an item for sale, eBay will charge you for the listing. Think of it like running a classified ad in the newspaper. Instead of being charged per word, like you are with a newspaper, you are charged according to your selling price.

eBay calls this basic charge their *Insertion Fee*.

The *insertion fee* for an item starting at less than .99 cents US might only be 20 cents. If an item is selling for more than $500 US, the fee could be $4.80. The fees vary a little with the times. And, the fees vary country to country.

Selling

You will be asked *and tempted* to add options to help sell your stuff. Be careful you don't add options worth more than what you are selling!

If your item sells, you are also charged a "Final Value Fee".

This is a percentage of the final bid. Currently, items under $25 are charged 5.25%. The percentage drops as the Final Value goes up.

Perhaps, the best advice I can offer, is to become familiar with eBay before you sell.

When you want to sell something, do a search for similar items. See how they are listed. Notice the category(s) the item is listed in and what similar things sell for.

eBay

Ready to sell?

Here's how:

1. Log on to eBay with your User ID.
2. eBay will ask you to register as a seller.
3. **You will be required to provide credit card information.**

Once you are a *Registered Seller*, these are the steps eBay will bring you through to list your item:

1. Choose a category(s) for your item.
2. Choose a title for your item.
3. Give details about your item, add a picture if you like.
4. State your payment and shipping terms.
5. Review. Make any changes. Submit!

Easy like pie!

Security

There are so many users on eBay, not because they want to make a quick buck, but because it's easy and fun.

But, not everyone in our world is honest. So, when you are surfing around, use common sense. If something looks quirky, it just might be.

Please, take time and look over the next couple of pages for a few simple tips, directly from eBay.

Knowing just a little can save you a lot!

DON'T BE SCAMMED BY EBAY IMPERSONATORS.

→ → →

The real thing is where it's at!

eBay

Security on and about ebay

Please read the following information. It has been **quoted directly from eBay.com.** It is important information to know.

"*E-mail and Web Sites Impersonating eBay*

- *eBay will never ask you to provide sign-in passwords, credit card numbers, or other sensitive information through e-mail. If we request information from you, we will always direct you back to the eBay site. With very few exceptions, you can submit the requested information through your "My eBay" page.*

- ***Take caution with e-mail that includes attachments or links***

- ***eBay will not send you e-mail that includes attachments*** *and you will not be required to enter information on a page that cannot be accessed from the eBay site. If you receive a message that appears to have been sent from eBay that includes an attachment, do not open it.*

Security on and about ebay

Continuing quote from ebay.com

Beware of fake Web sites pretending to be eBay

◇ *Only enter your eBay password on pages where the Web address (URL) begins with **https://signin.ebay.ca/**. Even if the Web address contains the word "eBay", it may not be an eBay Web page.*

◇ *These fake Web sites (also called "spoof" Web sites) try to imitate eBay in order to obtain your eBay password and access to your account.*

◇ *All genuine eBay.ca sign-in pages will begin with **https://signin.ebay.ca/**. Similarly, eBay.com sign-in pages will begin with **https://signin.ebay.com/**.*

◇ *Never type your eBay User ID and password on a page that doesn't have "ebay.ca" or "ebay.com" immediately before the first forward slash (/)."*

End quote from eBay.com

Bright Ideas

-
-
-
-
-
-
-

Instant Messaging

Instant Messaging, Abbreviated *IM*

Text messaging, instant messaging and chat rooms, seems like everyone is doing it! So, what are they?

- **Text Messaging** - small text messages sent between phones, often cell phones. Receiving text messages is usually free, but sending them costs a small amount. The fee is maybe around 15 cents. This wee amount can add up pretty fast, so if you are watching pennies, watch out!

- **Instant Messaging** - when you're using an IM program, you can type back and forth to friends in *real time!*

- **Chat Rooms** - Chat rooms are for groups of people with similar interests. You can instant message between the group, or post a message that people can respond to on a message board.

Instant Messaging

Is IM safe?

For your computer?
Mostly yes. I say "mostly" only because IM is so popular. There are many viruses written to infiltrate your system through it.

"A good defence is the best offence."

You need to have current anti-virus software and anti-spyware software running. Your IM program should also be up-to-date and you should keep your Microsoft Windows® updated too!

Most viruses are transmitted through attachments and *you have to open the attachment for it to be released.*

The best way to be safe from viruses is to delete messages with attachments!

Of course, friends will send you attachments. If you **know who sent the message** and feel it's clean and safe... Go ahead, OPEN IT!

Instant Messaging

Is IM safe for very young children?

NO, not without proper supervision and guidelines.

You would never drop a child off in the middle of Times Square without any supervision. Well, letting them go into IM without guidance is just about as bad!

You have to learn about internet safety, and you must teach your children how to use the internet safely.

Safety Tips and IM Etiquette

- Don't open attachments
- "Block" uninvited messages
- Only IM with people you personally know
- A Webcam - best used on a computer in the living room or kitchen - not a good idea in the privacy of a bedroom....
- IM addresses are NOT anonymous. They are traceable. It is illegal to make threats or illicit something bad. The police can and will track down illegal behavior.
- Understand that messages can be copied and forwarded on to other people. In that way emailing is NOT PRIVATE. Be careful what you say.

Instant Messaging

Safety & Etiquette

- People often change their "handle" (nickname) or will have more than one account in different names. Be sure you know who you are talking to.

- Feelings are easily hurt through IM, especially when kids are chatting back and forth. Something as simple as "Did you see what Mary was wearing today?", can turn into something pretty ugly.

- Think about what you are saying - and what you are not saying. Remember, the reader can't see your face or hear the inflection of your voice. Abbreviated text can easily be misunderstood!

- IM etiquette can be summed up by my Mom's simple rule. *Treat others with the respect and dignity that you expect to be treated yourself.*

Now that I've said all that...
Using IM is really kinda fun :-)
and a very efficient way to communicate!

Instant Messaging

How to start

Most PC's come with an Instant Messaging program pre-installed. You just have to open it, and then register with the service. Basic IM services are free. If you want something with more options, bells and whistles, you have to pay for it.

Here are some of the most popular instant messaging programs:

- AIM, from America On-line
- Windows Messenger, MSN (hotmail)
- Yahoo Messenger, from Yahoo

Signing up

The process is very easy. Deciding on a screen name or email name, can be the hardest part. Well, maybe not. Finding a name that is not already being used, that can be the hardest part!

The registration information is basic. You will be asked for your name, gender, birthdate, an alternate email address, and you will need to come up with a password to access your account.

Once a screen name has been accepted, you're in!

Instant Messaging

Signed up. Now what?
Let your friends know that you have IM and give them your IM e-mail address. Add your friends' addresses to your "contact" list.

You will find sending and receiving messages is very straightforward. If it wasn't, nobody would use it! Type a message and hit send! "Beep" you got one! Simple as that.

When you have your IM program open, you will receive messages from friends as soon as they send them. It is very much like using a telephone, but with text. Unlike your phone, you can have many conversations going on at once.

You can even have your program open and not accept messages, by changing your "status". Instead of being on-line, you can be: away, appear off-line, out-to-lunch... there's lots of options.

Every program has a different layout, but they all essentially DO the same things. Have a look through the options. Turn on or off sounds, set up what security levels you want, change your nick-name. Look around. Check out your options. Don't be scared!

Chat Rooms

When you text-message someone, or IM with someone, you are communicating one-on-one. When you use a "chat room" everybody in the *room* sees what you have to say.

IM programs have an option to create a personal chat room with your friends, great for making plans! If, during a conversation, you want to "invite" someone to join in, you can. Just have a look around for the invite option. In MSN it's under Actions > Invite contact into conversation.

Chat rooms are a great place for public discussion.

If I've got a problem with something, I might log into a group that deals with whatever it is. Perfect everyday strangers offering help and sharing lots of new ideas. It's really very cool.

If you "meet" someone in a **public chat room** and decide to get to know them better, be careful. You don't know the person. You only know what they have typed or told you, and it might not be true.

Never personally meet a stranger alone. Bring a friend, tell other friends what you are doing. Meet in public. Be street-smart and stay safe.

Text Talk

A New Language Is Born!
As text messaging grows, and devices get smaller, a language of acronyms has been born. Like it or not, it's probably here to stay.

Some acronyms are more common than others, some develop just between friends. You could even consider it "shorthand" for typers, often typed phonetically instead of correctly.

Text messages are cheap to send and are pretty non-intrusive to receive.
That's an improvement over someone taking a cell-phone call in the middle of talking to you!

To use your phone's key-pad to type out a word you have to "multi-tap" each number, until your letter comes up. For example:

for the letter A, press #2 once
for the letter R, press #7 three times
for the letter E, press #3 twice

Text Talk

Here is a list of a few acronyms and what they mean:

BF	boyfriend
B4	before
B4N	bye for now
BRB	be right back
CMON	come on
CU	see you
DIKU	do I know you
F2F	face to face
FYEO	for your eyes only
GF	girlfriend
GTG	got to go
IC	I see
LOL	laugh out loud
NO1	no one
NW	no way
OMG	oh my god
PM	private message
RL	real life
SUL	see you later
SUP	what's up?
TA	thanks alot
TC	take care
UR	you are
zzzz	sleeping (or bored)

Text Talk

Companies are trying to make texting easier.

Multi-tapping, pecking out letters for words, can be frustratingly slow. So, technologies are being developed to make texting easier. Here are a few of the new products out there, trying to help us out:

Predictive Text Programs
This program is often installed in new cell phones, it will try and guess what word you are tapping out before you finish it.

Fastap phones
These are phones that have placed individual letters around the number pad, so there is no need to multi-tap!

Treo's, Blackberry's, etc.,
Personal email devices have teeny-tiny complete keyboards to type out messages or letters. They come with a little poker to peck out the letters with! More than a phone or a daytimer, they can send and receive emails, take pictures and much more.

Personal Digital Assistants

PDA's - Blackberry's, Treo's etc.,

We used to call these things organizers or day-timers, but now-a-days they just do so much more. Some have so many functions they are practically a mini-laptop!

A couple of PDA's have the lions share of the market, the **Blackberry**, made by RIM and the **Treo**, made by the folks who gave us the Palm Pilot.

They are very popular devices in the business community, especially for people who work on the road a lot.

Like most technology, figuring out how to use it the first time can be frustrating; but once you get the hang of it, you might really enjoy the options these little PDA's offer!

Bluetooth technology

Go Wireless!
Nowadays, you will often hear that some new device has Bluetooth technology. OK. Great. What is it?

Bluetooth® is a wireless communications technology that allows devices to talk to each other. Imagine your computer system set up without any cables. Wow! With Bluetooth technology it can be.

One of the great things about this technology, is that many different manufacturers are using it. If a device has Bluetooth technology, it can be set up to work with almost any other device with Bluetooth.

For example, if you had a Xerox printer and Sony camera that both have Bluetooth technology, you could send your pictures directly to your printer without hooking anything up!

It might look funny when people are talking into thin air, but if they are wearing a Bluetooth earpiece, they are probably not talking to themselves! The earpiece is working together with a Bluetooth enabled phone!

Wireless is not science fiction any more!

Bright Ideas

-
-
-
-
-
-
-

Bright Ideas

-
-
-
-
-
-
-

Index

Index

Index

Index

Write your own index notes here.

Self publishing is a job with many hats. Author, Publisher, Marketer, Business Manager, Accountant... and I couldn't do it without the help of many talented people teaching me along the way.

Thanks to Halina St. James, (www.podiumcoaching.com) whose phenomenal expertise and advise taught me how to "Talk it out!"

To Dan Soucoup, Errol Sharp and Bev Rach for your publishing words of wisdom which manage to both excite and terrify me.

Thanks to Ellen Woodger of EPublicity who helped get My Parents First Computer out of the Maritimes and into the "big world." You helped it see a success that gave me the confidence to publish My Parents Second Computer and Internet Guide.

Thanks to Janet MacMillan of MTL for the introduction. Her talents and clear vision, help me navigate my way.

Thanks to all the amazing people of my community, who everyday, by example, give me the courage and confidence to tackle the unknown.

Thanks to Doreen Cassidy and my family of editors: Mom, Doug, Rosanne, Rob, Verna, Dan, Jane. Your help, comments, suggestions and corrections are always appreciated.

Last but not least, thanks to my husband and sons, for accepting the time and commitment being an entrepreneur takes, and encouraging me to go for it anyway.

"Thank you", to all of you, who kept asking for this; the second book in the *My Parents* series. It wouldn't have been written without you.

Your comments, suggestions, the answers to your questions, and that "just a little more" you asked for, is what you will find between these covers.

Easy is good. So, *Beyond the Basics* offers the same simple, just-do-it style, found in *My Parents First Computer and Internet Guide.*

Learning computer basics - and beyond - can be easy and in the *My Parents* series, I make sure of that!

Sales from this book will continue to help me raise funds for the Canadian Cancer Society.

Find a cure.

Your comments are always welcome,
visit me at www.myparentsfirst.com